Akshay Kumar

Power Improvement of DFIG Wind Turbine Systems Connected to Grid using Buck Boost Convertor

GRIN Verlag

Bibliografische Information der Deutschen Nationalbibliothek:

Die Deutsche Bibliothek verzeichnet diese Publikation in der Deutschen National-
bibliografie; detaillierte bibliografische Daten sind im Internet über http://dnb.d-
nb.de/ abrufbar.

Imprint:

Copyright © 2014 GRIN Verlag GmbH
Druck und Bindung: Books on Demand GmbH, Norderstedt Germany
ISBN: 978-3-656-69809-8

This book at GRIN:

http://www.grin.com/en/e-book/276637/power-improvement-of-dfig-wind-turbine-
systems-connected-to-grid-using

GRIN - Your knowledge has value

Der GRIN Verlag publiziert seit 1998 wissenschaftliche Arbeiten von Studenten, Hochschullehrern und anderen Akademikern als eBook und gedrucktes Buch. Die Verlagswebsite www.grin.com ist die ideale Plattform zur Veröffentlichung von Hausarbeiten, Abschlussarbeiten, wissenschaftlichen Aufsätzen, Dissertationen und Fachbüchern.

Visit us on the internet:

http://www.grin.com/

http://www.facebook.com/grincom

http://www.twitter.com/grin_com

Power Improvement of DFIG Wind Turbine Systems Connected to Grid using Buck Boost Convertor

Akshay kumar
Dept. Electrical and Electronics Engineering,
AKGEC Ghaziabad, INDIA.

Abstract- **This paper proposes an application of Buck booster enhancement which is capable of smoothening the power of doubly-fed induction generator (DFIG) wind turbine systems feed to the grid. A grid-side converter (GSC) is used to preserve the DC-link voltage. Other side, a rotor–side converter (RSC) is used to maintain the active and reactive powers of DFIG-Wind Turbine system. This analysis is made to analyze the active power sharing between the DFIG and the grid. This whole system is simulated in the MATLAB Simulink environment.**

Index Terms— Doubly fed induction generator (DFIG), DC-link voltage control, Buck booster, wind energy conversion system (WECS), Power smoothening.

I. INTRODUCTION

Fossil fuels decreasing day-by-day and their side-effects harm humans, heritages monuments, etc. So, renewable energy has become an important energy source. Wind energy is the rapidly growing and most promising renewable energy source among other due to economically point of view [1].

Wind power will be adequate of becoming a major contributor to the electricity supply in India. However, there are various challenges associated with designing such power systems, which is generated from the wind power.

As we know that the commercial DFIG wind turbines using the technology that was developed around 1980's, based on the traditional decoupled *d-q* vector control technique. This paper shows that the limitation in the older vector control approach used for the grid-side converter of the DFIG wind turbine. This blames a distinct linear modulation limit when it is used beyond its limit. This type of limitation has also been described in details by researchers [2],[3].

This paper forms an improved control of DFIG wind turbine under a direct-current vector control

configuration using Buck-boost convertor. A steady state simulation of DFIG-WECS is necessary to understand the behavior of WECS using MATLAB Simulator and several characteristics of DFIG are analyzed.

II. WIND TURBINE MODELING

Wind turbine converts the kinetic energy of wind into mechanical energy for producing the torque. The energy includes the wind is in the form of kinetic energy, its magnitude depends on the wind velocity and the air density. This wind power developed by the turbine is shown in the below equation [4]:

$$P_m = \frac{1}{2} c_p(\lambda, \beta) \rho \, A v^3 \qquad (1)$$

Where c_p is the Power Co-efficient, A is the area of the turbine blades (in m^2), ρ is the air density (in kg/m^3) and V is the wind velocity (in m/sec). The power coefficient (c_p) is defined as the power output of the wind turbine to the available power in the wind regime. This coefficient determines the "maximum power" the wind turbine can consume the available wind power at a given rated wind speed. It is a function of the tip-speed ratio (λ) and the blade pitch angle (β). The blade pitch angle can be controlled by using a the tip-speed ratio (TSR) and "pitch-controller" is given as [5]

$$\lambda = \frac{\omega R}{v} \qquad (2)$$

Where ω is the rotational speed of the generator and R is radius of the rotor blades.

So, TSR can be controlled by controlling the rotor speed of the generator. For a given rated wind speed, there is only one rotational speed of the wind generator which gives an extreme value of c_p, at a given β. This is the main principle behind "maximum-power point tracking" (MPPT) [5], [9] and structure of wind turbine is made by taking this principle in consideration.

III. DOUBLY FED INDUCTION GENERATOR

The wind turbine with doubly-fed induction generator (WTDFIG) is shown in the fig.1. The AC/DC/AC converter is divided into two components- the rotor-side converter (C_{rotor}) and the grid-side converter (C_{grid}), and buck boost between them. C_{rotor} and C_{grid} are using the forced-commutated power electronic devices (mainly IGBTs) to synthesize an AC voltage from a DC voltage source. A capacitor connected on the DC side for smoothing the supply. A filter is used to connect GSC (C_{grid}) to the grid. The three-phase rotor winding is connected to RSC (C_{rotor}) directly through slip rings and brushes, and the three-phase stator winding is directly connected to the grid. The power captured by the wind turbine is converted into electrical power by the induction generator and it is transmitted to the grid by the stator and the rotor windings. The control system generates the pitch angle control signal and the voltage control signals for C_{rotor} and C_{grid} respectively. In order to control the power of the wind turbine [6].

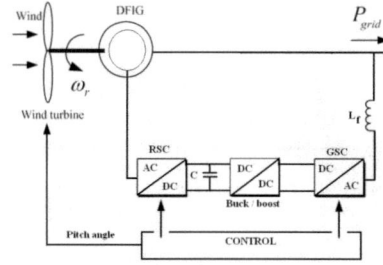

Fig.1. The Wind Turbine and the Doubly-Fed Induction Generator System.

IV. OPERATING PRINCIPLE OF THE WIND TURBINE DOUBLY-FED INDUCTION GENERATOR

Fig.2 demonstrates the power flow of DFIG, which is used to define the operating principle. In the figure of power flow, followings parameters are used:

P_m Mechanical power captured by the wind turbine and transmitted to the rotor

P_s Stator electrical power output

Q_s Stator reactive power output

P_r Rotor electrical power output

Q_r Rotor reactive power output

P_{gc} C_{grid} electrical power output

Q_{gc} C_{grid} reactive power output

ω_r Rotational speed of rotor

T_m Mechanical torque applied to rotor

T_{em} Electromagnetic torque applied to the rotor by the generator.

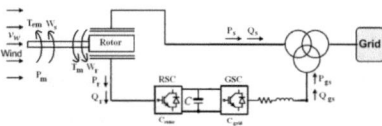

Fig.2. The Power Flow

Rotational speed (ω_s) of the magnetic flux in the air-gap of the generator is called synchronous speed. It is directly proportional to the number of generator poles and to the frequency of the grid voltage, combining rotor and wind turbine inertia coefficient.

The mechanical power and the stator electric power output are computed as follows [6]:

$$P_m = T_m \omega_r \qquad (3)$$

$$P_s = T_{em} \omega_s \qquad (4)$$

For a loss-less generator, the mechanical equation is

$$J \frac{d\omega_r}{dt} = T_m - T_{em} \qquad (5)$$

In steady-state at fixed speed for a loss less generator

$$T_m = T_{em} \ \& \ P_m = P_s + P_r \qquad (6)$$

It follows that:

$$P_r = P_m - P_s = T_m \omega_r - T_{em} \omega_s$$
$$= -T_m \left(\frac{\omega_s - \omega_r}{\omega_s} \right) \omega_s = -s T_m \omega_s$$
$$= -s P_s \qquad (7)$$

where s is defined as the slip of the generator:

$$s = \left(\frac{\omega_s - \omega_r}{\omega_s} \right) \qquad (8)$$

V. ROTOR CONTROL SYSTEM

RSC is used to control the wind turbine output power and mainly the voltage (or reactive power) measured at the grid. The power is controlled in order to track a pre-defined power-speed characteristic. An example of such a characteristic is shown in the fig.3 called Turbine Characteristics and Tracking Characteristic, by the ABCDE curve superimposed to the mechanical power characteristics of the turbine obtained at different wind speeds. The actual speed of the turbine ω_r is measured and the corresponding mechanical power of the tracking characteristic is used as the reference power for the power control loop. The tracking characteristic is defined by five points: A, B, C, D and E. From zero speed to speed of point A the reference power is zero. Between point A and point B the tracking characteristic is a straight line, the speed of point B must be greater than the speed of point A. Between point B and point C the tracking characteristic is the locus of the maximum power of the turbine. The tracking characteristic is a straight line from point C and point D. The power at point D is one per unit (1 pu) and the speed of the point D must be greater than the speed of point C. At and beyond point E, the reference power is a constant equal to one per unit (1 pu).

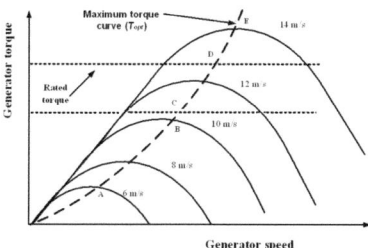

Fig.3. Wind Turbine Characteristics and Tracking Characteristic

The standard power control loop is shown in the fig.4 called RSC Control System. The real electrical output power of wind turbine is measured at the grid terminals, added to the total power losses (mechanical and electrical) and is compared with the reference power. To reduce the power error to zero, a proportional-integral (PI) is used. PI output is the reference rotor current I_{rqref} that must be injected in the rotor by RSC. This is the current component that produces the electromagnetic torque T_{em}. The real I_{rq} component of positive-sequence current is compared with I_{rqref} and current regulator PI is used to reduce the error to zero. The output of this PI is the voltage V_{qrcs} generated by RSC. The current regulator is supported to feed forward terms which forecast V_{qrcs}.

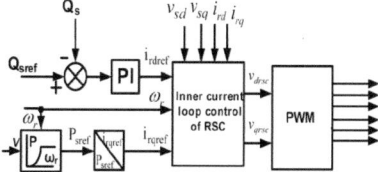

Fig.4. Block diagram of Rotor-Side Converter Control System

The voltage or the reactive power at grid terminals is controlled by the reactive current flowing in the RSC. When the wind turbine is operated in voltage regulation mode, it shows the following V-I characteristic [6] shown in fig.5.

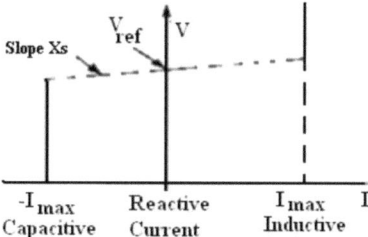

Fig.5. Wind Turbine V-I Characteristic

VI. GRID CONTROL SYSTEM

GSC is used to regulate the voltage of the DC bus, given by buck boost convertor. This model allows GSC to generate or absorb reactive power. The control system is shown in the fig.6 called Grid-Side Converter Control System, consists of [7]:

1. Current components abc of AC are converted to d and q components to control the positive sequence current and control the DC voltage.
2. DC regulator generates the reference current for the current regulator.
3. Current regulator controls the amplitude of voltage generated by GSC as well as controls the phase of the voltage by using the reference current generated by the DC voltage regulator.

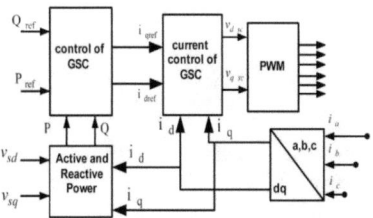

Fig.6. Grid-Side Converter Control System

VII. PITCH ANGLE CONTROL SYSTEM

The pitch angle is retained constant at zero degree until the speed reaches point D speed of the tracking characteristic. At or beyond point E the pitch angle is proportional to the speed deviation from point E speed. The control system [8] is shown in the following fig.7.

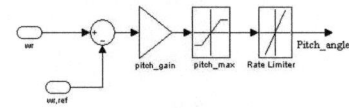

Fig.7. Pitch Control System

VIII. CONFIGURATION AND CONTROL OF BUCK BOOST

Fig.8 shows the configuration of buck boost as well as the control, this block is connected between the rotor side convertor and grid side convertor. Buck boost convertor contains IGBT with a battery. The operation state depends upon the supply power, whether it operates in buck mode or in boost mode.

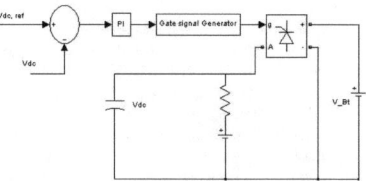

Fig.8. Buck / boost convertor

By comparing the dc link voltage with reference voltage, a gate pulse is generated according to the compression of voltages as shown in above figure. This method is verified in the matlab-simulation.

XI. MATLAB-BASED MODELING

The model of DFIG-WECS with Buck boost, shown in Fig.1 is developed in the MATLAB-SIMULINK shown in fig.9 and results are presented to demonstrate its behavior.

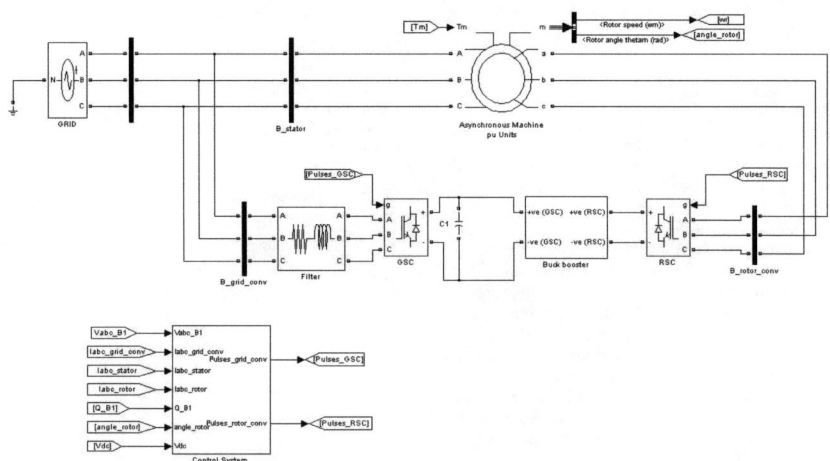

Fig.9. GSC and RSC control of DFIG wind turbine in Sim-Power Systems

Performance of the proposed configuration of a DFIG-based WECS is shown in fig.10. The waveforms for grid voltage, grid current, grid side converter current, rotor side converter current, stator current, rotor speed, and dc link voltage are presented. The convention for the buck boost battery power is chosen as to be negative if the battery giving (discharges) any power to the grid and positive if power is stored in the battery of buck boost.

Power of grid is maintained to be constant at 0.65 MW by the proposed grid power control strategy. But, this is retained by either charging or discharging the battery of buck boost in the corresponding region of operation. The reactive power is retained at a stable value of zero, demonstrating a unity power factor operation. This study has been performed at variable wind speeds and the grid power is retained to be constant at the reference value. Hence, the grid power reference is chosen to be 0.65 MW as calculated and satisfactory results are obtained, as shown in Fig.10.

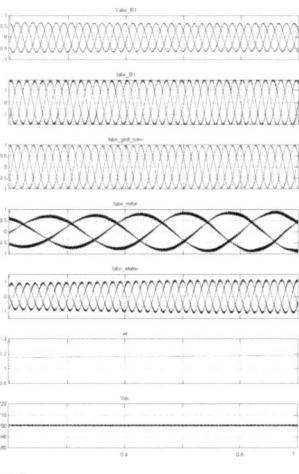

Fig.10. Performance of a DFIG-based WECS with a Buck boost.

X. CONCLUSION

A configuration of a DFIG-based WECS with a Buck boost in the dc link has been proposed with a control strategy to maintain the grid power constant. A methodology to design the Buck boost has been proposed by considering practical data.

The performance of the proposed control strategy on a DFIG-based WECS with Buck boost has been demonstrated under different wind speeds. It has been observed that DFIG-based WECS with Buck boost demonstrates satisfactory performance under different wind speed conditions. If the utility fails to maintain the grid power constant, then during periods of "over-generation," the consumers are to be paid in return to implement "load-leveling," and absorb the excess power.

This is an unbeneficial practice as the supplier looses both energy and money. The proposed configuration and control strategy mitigates this, by supplying a constant power to the grid throughout and thus retaining a constant flow of energy to the grid irrespective of the variations in the wind speed. Placing a Buck boost in the dc link of a DFIG-based WECS proves to be a satisfactory implementation in terms of maintaining a constant power at the grid.

APPENDIX

A. Parameters of the Wind Turbine

Parameters	Value
Rated Power	1.5Mw
Cut-in Wind speed	8m/s
Rated Wind speed	12m/s
Cut-out Wind	16m/s
No. of Blades	3
Rotor Diameter	82m
Swept Area	5281m^2

B. Parameters of the buck boost Battery

Parameters	Value
Battery Nominal Voltage (DC-link)	1200 V
DC-link Capacitor	5000 μF
IGBT battery voltage	700 V
Battery Series Resistance	0.094 μΩ

C. Parameters of the DFIG

Parameter	Value
Rated Power(MW)	1.5
Stator Voltage(V)	575
Frequency(Hz)	50
Pole numbers	6
Stator Resistance(pu)	0.00706
Rotor Resistance(pu)	0.005
Stator leakage Inductance(pu)	0.171
Rotor leakage Inductance(pu)	0.156
Magnetizing Inductance(pu)	0.3
Inertia constant(s)	5.04
Friction Factor(pu)	0.38

REFERENCES

[1] A. Miller, E. Muljadi, and D. Zinger, "A variable speed wind turbine power control," *IEEE Trans. Energy Convers.*, vol. 12, no. 2, pp. 181–186, Jun. 1997

[2] R. Pena, J. C. Clare, and G. M. Asher, "A doubly fed induction generator using back to back PWM converters supplying an isolated load from a variable speed wind turbine," in *Proc. Inst. Elect. Eng., Electr. Power Appl.*, May 1996, vol. 143, no. 3, pp. 231–241

[3] L. Xu and W. Cheng, "Torque and reactive power control of a doubly fed induction machine by position sensorless scheme," *IEEE Trans. Ind. Appl.*, vol. 31, no. 3, pp. 636–642, May/Jun. 1995.

[4] G. M. Masters, "Renewable and Efficient Electric Power Systems," Hoboken, NJ: *IEEE Press*, Wiley-Interscience, 2004

[5] S. Sathana and Ms. Bindukala M.P , "Hybrid Solar and Wind Power Conversion Using DFIG with Grid Power Leveling," *International Journal of Emerging Trends in Electrical and Electronics (IJETEE)* Vol. 1, issue. 1, March-2013

[6] Mohamed Amer Hassn Abomahdi, A. K. B hardwaj, Surya Prakash and Mohammad Tariq, "Commanding Doubly-Fed Induction Generator (DFIG) to Decouple Active and Reactive Power for wind energy," *IOSR Journal of Electrical and Electronics Engineering (IOSR-JEEE)* e-ISSN: 2278-1676,p-ISSN: 2320-3331, Volume 7, Issue 3 (Sep. - Oct. 2013), PP 12-19

[7] Richard Gagnon, Gilbert Sybille, Serge Bernard, Daniel Paré, Silvano Casoria and Christian Larose, "Modeling and Real-Time Simulation of a Doubly-Fed Induction Generator Driven by a Wind Turbine," *International Conference on Power Systems Transients (IPST'05)* in Montreal, Canada on June 19-23, 2005 Paper No.IPST05-162

[8] Branislav Dosijanoski, "simulation of doubly fed induction generator in a wind turbine," *XI international Phd workshop* owd 2009, 17-2- october 2009

[9] Vijay Chand Ganti, Bhim Singh, *Fellow, IEEE*, Shiv Kumar Aggarwal, and Tara Chandra Kandpal, "DFIG-Based Wind Power Conversion with Grid Power Leveling for Reduced Gusts," *IEEE Transactions On Sustainable Energy*, vol. 3, No. 1, January, 2012.

Akshay Kumar was born in Ghaziabad, India, in 1988. He received the B. Tech (Electrical and Electronics) degree from Ideal Institute of Technology, Ghaziabad (affiliated to UPTU, Uttar Pradesh), in 2009 and, currently pursuing his M. Tech (Electrical Power and Energy Systems) from Ajay Kumar Garg Engineering College, Ghaziabad (affiliated to MTU, Uttar Pradesh). He worked in Industry and he has also teaching experience of more than 1 and half year in Engineering College.